I0441345

Table of Contents

List of Figures

List of Tables

Introduction

In today's world, thousands of companies are developing their own software. Others are purchasing products off the shelf and customizing them to fit their needs. In these cases, who will write the instructions to show the employees how to use the software? Who will conduct the training?

How to Write In-House Software User Manuals was written to help you create a simple or complex user manual. The instructions are easy to follow, and anyone with intermediate computer knowledge can put together a user manual for any type of software.

This book details the specifics of using Microsoft Word® to produce a software user manual. It also demonstrates how to use SnagIt™, an imaging application, to capture and save software screen shots and to insert the screen shots into a Word document.

You may already know how to perform some of the instructions in this book, but I have covered all the bases from beginning to end. If you come to a section that you have experience in and can perform the steps, skip the section and move to the next one.

The design and writing instructions in *How to Write In-House Software User Manuals* are very basic. However, the steps are not set in stone. You can modify any of the instructions to your preference. If you do make your own changes, be certain that the company that owns the rights to the manual is in agreement. It is a smart idea to let the manager, supervisor, or person in charge of the project approve the format of the manual after the first section or two are completed. Doing so may save you valuable time in the event that the manager does not like the format and wants to make changes.

How to Write In-House Software User Manuals shows you how to:

- Gather the necessary information before you begin writing.
- Create a cover page that includes the company logo.
- Outline the manual, starting with the Table of Contents.
- Create headings using the "how to" approach.
- Write step-by-step instructions.
- Set conventions using different formats, colors, and fonts.
- Create screen shots using SnagIt.
- Format the document, including headers and footers.
- Insert a Table of Contents generated by Word.
- Insert a List of Figures generated by Word.
- Insert a List of Tables generated by Word.

As a bonus, *How to Write In-House Software User Manuals* contains an appendix that offers a refresher of simple grammar rules, punctuation rules, and technical writing rules. It does not contain *all* the rules; but it does include a few that you may encounter during the writing process of your manual. After all, there are so many rules, we all need a little help now and then to refresh our memory.

Gather Information

Diving head first into a writing project before you know the essentials makes writing the manual difficult and could possibly delay deadlines. Steps taken at the beginning of the project are just as important as the actual writing of the manual. Before the writing can begin, there are many factors you need to know so that the project can flow smoothly.

This section discusses the things you need to do to learn all the particulars and issues of the project before you begin the writing process.

Schedule a Meeting

First, schedule a kick-off meeting with everyone who is involved in the project. Take plenty of paper and a pen. Write down any observations or statements made by the attendees. In addition, make a list of the following:

- Company's physical location (especially if the project participants are located in separate divisions)
- Audience for the manual (Will the readers be mid-level employees, managers, corporate personnel, or other?)
- Subject Matter Expert (SME) names
- Deadline of the project

It is also necessary that you know who is in total charge of the overall project in the event that a disputed issue arises during the writing of the manual.

Study the Software

Before you start writing the manual, get access to the software you will be writing about. If you have never used or seen the software before, this is a good way to get an insight into how the users will see it for the first time. Play with the software to get familiar with its features and functions. Go from screen to screen, and pay close attention to tabs, fields, buttons, options, and any other information on the screen. Also note any fields that are automatically populated by the software.

While going through the software, jot down words or actions that could be used as headings or subheadings in the manual. This will help you tremendously when you get ready to create the manual outline.

If the company has any written material on the software, whether it is completed or sketchy notes, review the information. Make notes on what needs to be written more clearly or in more detail.

Next, make a list of questions for the SMEs. Is something confusing to you? If so, it would also be confusing to the reader. Should some of the options on a particular screen be rearranged to make logical sense? Should a cancel option be included on a screen in case the user makes an error and wants to go back to the previous screen or start over? Be sure to write down ALL questions that occur to you, no matter how irrelevant you think they may be.

Prepare for the Interview

Now that you have familiarized yourself with the software, the next step is to interview the SMEs. An SME could be the person who actually invented the software or the expert who has used the software for quite some time. Or it could be both.

Before beginning the interview, add the following information to your list of SME's names:

- Office and cell phone numbers
- Email addresses
- Physical location(s) if at a different location
- Section(s) of the software where the SME has the most experience

- SME's work schedule for the time it will take to complete the manual so that you know if they have planned a vacation or time off during the project or if they have a business trip scheduled

Keep the above information on hand at all times during the writing process. Sometimes you need answers or guidance, and it cannot wait. If you have difficulty getting help, insert a placeholder in the document with the question or a description of the problem and who is responsible for providing the information. Highlight the placeholder so that it draws your attention each time you review the manual. Make every attempt to get the information as soon as possible so that it does not prevent you from meeting the project's deadline.

If possible, get the name of a backup SME in case the first SME is unavailable at a critical time when you are writing the manual. Email the backup SME's answers or suggestions to the original SME so that both parties are in agreement of the information provided.

Interview the SME

Now it is time to actually sit down and interview the SMEs. Sit at a desk or table with the software running on a computer. This helps you to show the SME the screen or action in question, and it helps the SME explain the answer more thoroughly.

After interviewing all the SMEs involved, type your notes in a separate document. Group the notes that are related to one another. If you cannot remember something the SME said or pointed out in the software, do not be afraid to go back and ask again. On occasion, communication is misinterpreted or misunderstood. It is perfectly acceptable for you to ask a question more than once. After all, you did not write the software. You are learning it just as the users will learn it. If a concept is difficult for you to understand the first time it is explained to you, it might take some users several times to get it right. In that case, a different approach to the subject may be necessary.

Create the Cover Page

Now you can begin creating the manual. The first step is to create the cover page. Typically, the cover page contains the:

• Company logo

• Software name

• Type of manual (user, field reference, guidelines)

• Version (software or manual version)

• Confidentiality statement (optional)

• Copyright information (optional)

Keep in mind that the manager of the project may have his or her own ideas on how the cover page will look. Perhaps he or she may want the copyright information to stand on a page of its own. There is also a possibility that the company has documentation standards that you must follow. Whether you work directly for the company or are performing the work as a contractor, apply all standards to the formatting required by the company. If acceptable, you may apply some ideas of your own.

Most major corporations have specific logos that are used for documentation. Be sure that you use correct logos. The same goes for the confidentiality statement. This may be required by the company's legal department and must be worded in a specific manner.

The cover page can also contain any other information that is important to the publication of the manual. It is wise to get approval from several managers or supervisors so you can be sure that you have included everything that is needed on the page.

Outline the Manual

Now that you have finished interviewing the SME, organized your notes, and created the cover page, it is time to make an outline of the manual. The first section is the Table of Contents. Then, if applicable, include a List of Figures and a List of Tables section. This is called the "front matter" of the manual.

The outline flow of information in the manual is critical for the success of the manual. For instance, just because the software takes you from one screen to the next does not necessarily mean that the user will take that exact path when entering information into the software.

On the other hand, there are some software manuals that require the user to enter information into the first screen before the application allows him or her to advance to the next. Also, in some instances, the information the user enters into the first screen determines what he or she sees when he or she clicks to the next screen. Or the next screen. Or four screens later. Some software applications automatically populate fields with information related to other fields, performing instant calculations.

Therefore, take into consideration the end result of the information entered into the application and go from there. Is it strictly for reports? Does the end result determine if a job is feasible? Would the company lose money if they pursued the project? Is the software used only as a database and the flow of the outline is not an issue?

Take a look at how I started this book. I wrote an introduction first, explaining the details covered in the book. Next, I created the section *Gather Information*. Complete this section before you begin writing the instructions because you first need to know the SMEs assigned to the project, the intended audience, the deadline, and other major aspects of the project.

When creating names for your headings, insert a invisible "how to" before the main headings. For example, *(How to) Gather Information.* Or *(How to) Create the Cover Page.* The latter example is simply *Create the Cover Page.* This method can also be used for subheadings. After all, a software user manual is filled with "how to" instructions.

Review the Table of Contents of this book and notice how I worked the outline in the order of producing a book. This gives you an example to go by. But notice that the instructions on creating the front matter of a manual are at the end of this book. The headings, figures, and tables within the text must be created first and inserted into the manual before Word can generate the lists. Wait until the manual is finished—or nearly finished—and then generate the lists.

Create Headings

In a sense, writing the actual, step-by-step instructions is the easy part. Formatting the text takes more time. Even though the instructions are simple and easy to follow, the appearance of the information as a whole is extremely important. The users must *want* to read the instructions. If the information is too cramped and the pages contain limited white space, the user may become frustrated and not read it at all.

Color, graphics, and tables are all essential for categorizing information, and margins are just as important. Look closely at the headings of this e-book. They are a golden color and set apart from the body of the information. You can easily look down the left side of a page and quickly locate the headings. That is why I indented the body and step-by-step instructions further to the right. I also centered the screen shots according to the body margins, not the margins of the document itself.

The headings of each section or chapter must be very visible to the reader. In order for Word to generate a Table of Contents, you must apply a "style" to the heading. Attach the style "Heading 1" to the main headings. The instructions written in this section explain how I created the headings in this book and applied the styles.

Follow these steps when applying a style to a heading. These same steps can be used when applying styles to subheadings, paragraphs, or other forms of information.

1. Type the heading and leave your cursor on the paragraph line.

2. Select **Format > Styles and Formatting**.

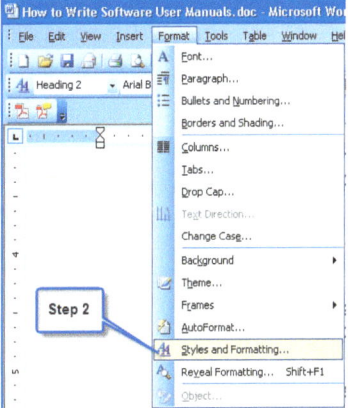

NOTE: The Styles and Formatting screen appears to the right of your document. Use this screen to create or modify the styles.

3. Select **Heading 1** in the *Styles and Formatting* screen.

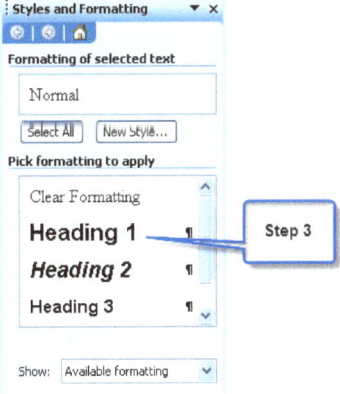

NOTE: The display in the Styles and Formatting screen show the current font and size of the styles. When you select a particular style, it appears at the top field where "normal" is displayed in the above illustration.

4. Move the cursor to the right of **Heading 1** and click the down arrow.

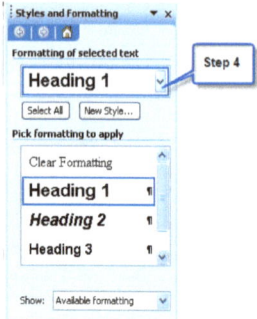

5. Select **Modify**.

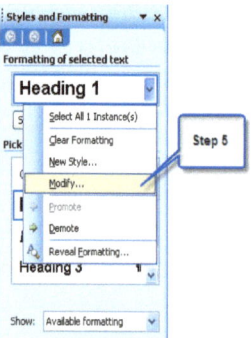

6. Click the down arrow in the **Font Name** field and select **Arial** in the *Modify Style* screen.

Create Headings

7. Click the down arrow in the **Font Size** field and select **14**.

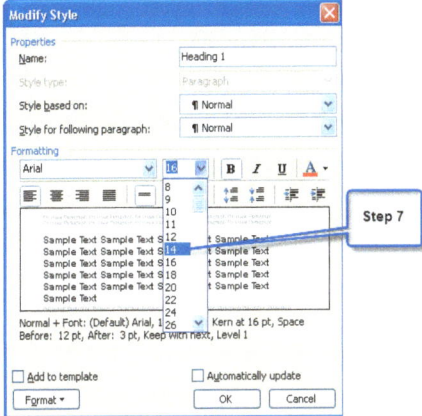

NOTE: Make sure the "bold" button is selected for all levels of headings for easy locating.

8. Click the **Format** button at the bottom left and select **Font**.

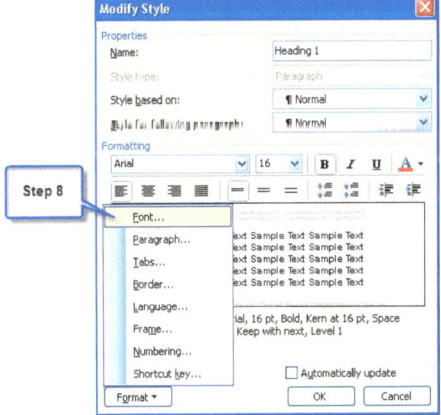

9. Click the **Font color** down arrow and select **More Colors** in the *Font* screen.

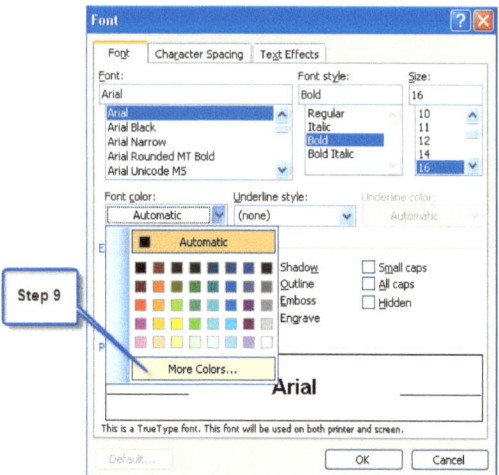

NOTE: You can also perform steps 6 and 7 in this screen.

10. Click the **Custom** tab on the *Colors* screen.

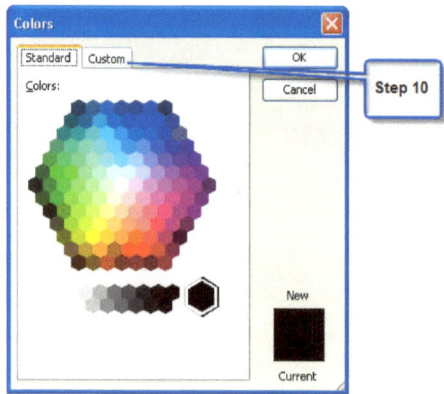

11. Type **187** in the **Red** field, **148** in the **Green** field, and **39** in the **Blue** field.

12. Click the **OK** button.

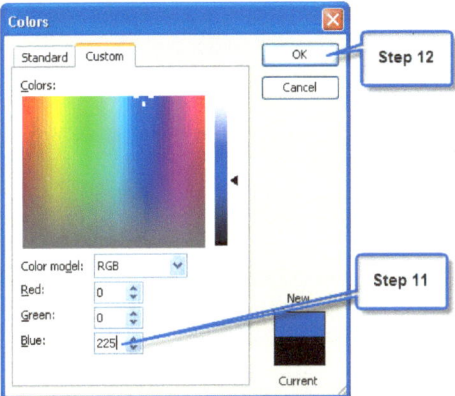

NOTE: Notice that the heading you created is exactly the same color as the headings in this book. These numbers combined recreated the color. Doing this ensures that you can match up colors perfectly for different styles.

13. Click the **OK** button in the *Font* screen.

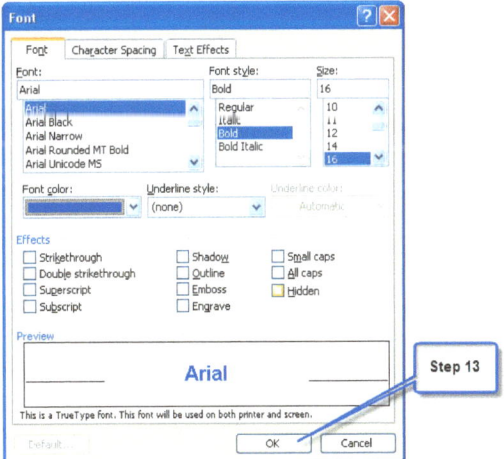

14. Click inside the **Automatically update** box in the *Modify Style* screen.

15. Click the **OK** button.

NOTE: Clicking the Automatically update button makes the latest changes to all items tagged with this style.

Notice that the Heading 1 style in the *Styles and Formatting* screen is now gold in color and bold.

Subheadings

Subheading styles are applied in the same manner as Heading 1. Use the Heading 2 style for subheadings and Heading 3 for sub-subheadings. The rule is to NOT go deeper than Heading 3. Too many subheadings cause confusion for the reader.

When applying styles to subheadings, use a font size two sizes smaller than Heading 1. This distinguishes the difference between Heading 1 and Heading 2. If you use Heading 3, apply an italic style or underline style to differentiate it from Heading 1 and Heading 2.

You may want to indent Heading 2 just a little farther to the right than Heading 1. However, I used the same margins for both headings in this book. This helps to save space.

You can include Heading 3 in the Table of Contents, but sometimes this makes the list too long and complex. Use your own judgment when deciding if you should including Heading 3 in the Table of Contents.

Write Step-by-Step Instructions

Writing step-by-step instructions can be very tedious. When writing instructions about a particular software screen, examine the screen closely. If possible, write the steps in the order the user will use them on the screen itself. Some software requires the user to jump around on the screen when filling in information. If this is the case, take your time. Think about how to put the instructions in a logical order. Then write the steps. Next, perform the steps, and see if they are clearly written or should be rearranged for better organization.

Number all steps within a group, starting at 1. When you start writing steps for a different section of the manual, remember to start the steps over at 1. The easiest way to do this is to apply a numbering style to the steps.

The steps below show you how to apply a numbering style to a set of instructions.

1. Type the information for the first and second steps and hit the **Enter** key.

Select·**Tools·>·Template·Add-ins**·on·the·toolbar.¶
Click·the·**Organizer**·button·on·the·*Template·and·Add-ins*·screen.¶

2. Select **Format > Styles and Formatting**.

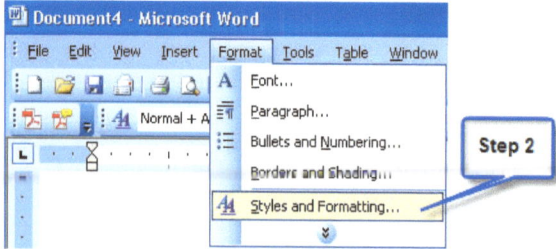

3. Click the **New Style** button in the *Styles and Formatting* screen.

4. Type the words **Numbered Steps** in the **Name** field in the *New Style* screen.

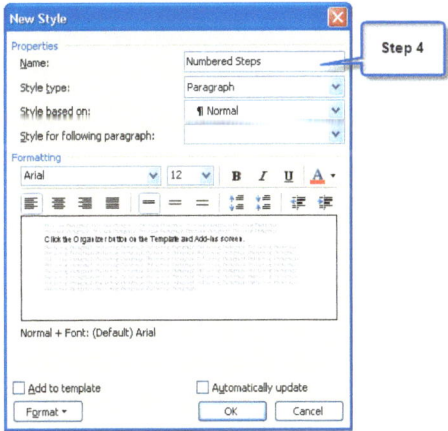

NOTE: For these instructions, I used "numbered steps" as the style name. However, you can choose whatever name you like, just as long as you can easily remember the style when you search for it in the style list.

5. Click the **Format** button and select **Numbering**.

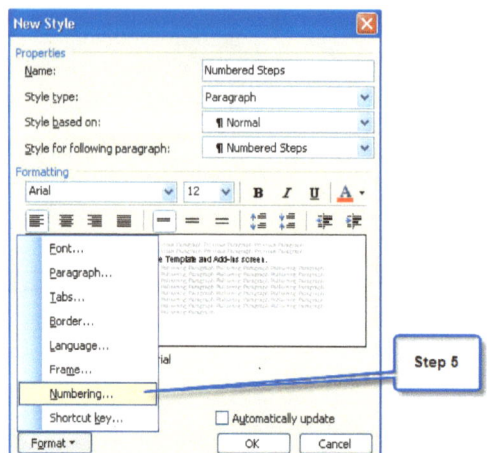

6. Select the **Numbered** tab, and then click one of the styles on the *Bullets and Numbering* screen that best fits the numbering characters you prefer.

7. Click the **Customize** button.

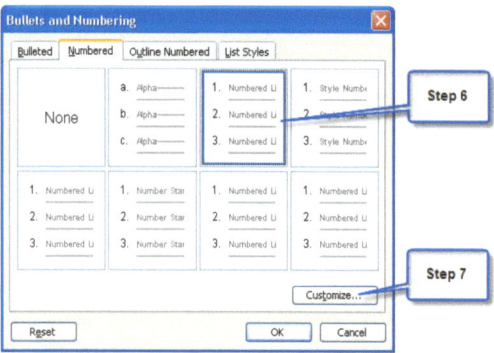

8. Check the **Number format** field on the *Customize Numbered List* screen to make certain the accurate number format is displayed.

   NOTE: Also check the **Start At** *field on the screen. The number "1" should appear in the field. Change the information in any of the remaining fields to fit the style you want.*

9. Click the **OK** button.

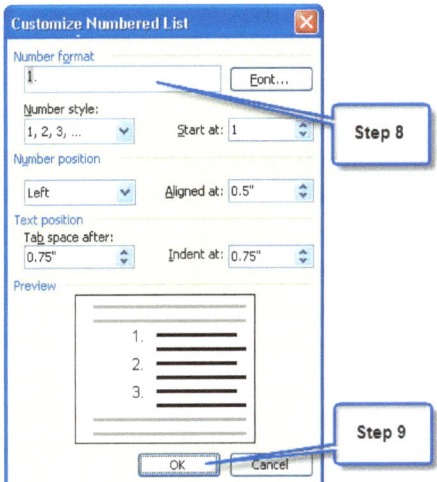

NOTE: You can also set the style margins within this screen.

10. Click the **Automatically update** box on the *New Style* screen.

11. Click the **OK** button.

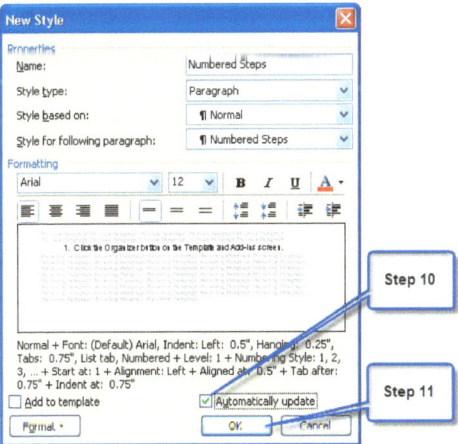

NOTE: The new style Numbered Steps appears under the Styles and Formatting screen to the right. It also has a "1." in front of the style indicating that the style is numbered.

12. Go back to your document and select the first two steps you typed.

Select **Tools > Template Add-ins** at the top of the Word document screen.
Click the **Organizer** button on the *Template and Add-ins* screen.

13. Move your cursor to the right of the document and select the style **1. Numbered Steps** in the *Styles and Formatting* screen.

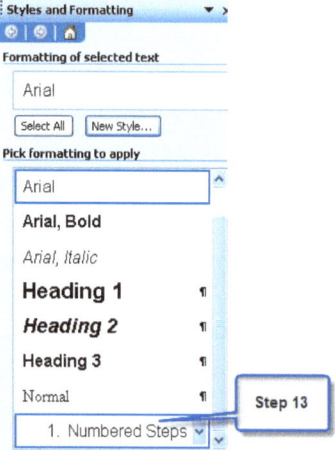

The steps now have a "style" applied to them. To type in step 3, simply hit return after the information in step 2. The next line is automatically numbered.

Some software manuals require an alpha list. In this case, create another style with the name "Alpha." When applying a numbering system, select the down arrow in the **Number style** field on the *Customize* screen and select a, b, c.

Renumbering Steps

I realize that most of you working with Word every day know how to apply styles and change the formatting. However, I have come across some strange experiences when using Word. When you start a new section of numbered steps and apply the style you have created, you need to make a few extra clicks to get the numbering to start back at "1."

Follow these steps when inserting a new group of numbering.

1. Select the first line of numbering.

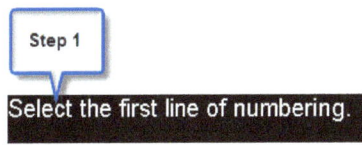

2. Select the appropriate numbering style in the *Styles and Formatting* screen.

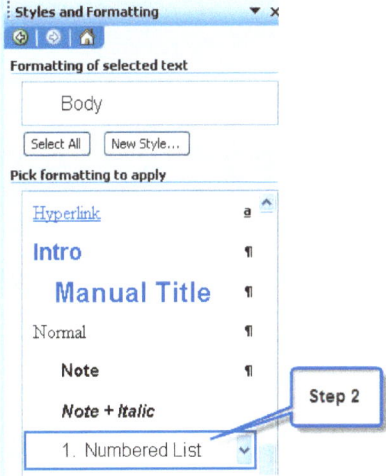

NOTE: Sometimes the numbering continues from the previous number list. For instance, the numbering may begin with "14" if the previous list contains 13 steps.

3. Right-click on the first numbered step.

4. Select **Restart Numbering**.

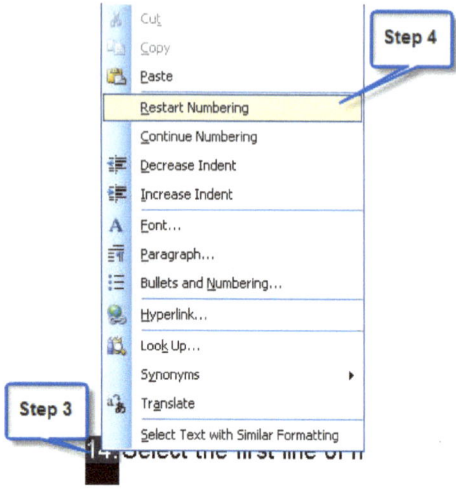

Now the first step begins with 1.

CAUTION: Sometimes the renumbering does not work even if you follow the above steps. When this happens, cut the steps from your current document and paste them into a blank document. Now the steps begin with 1. Press Ctrl > A to select all items in the new document. Copy and paste them back into the working document. The numbering should be fixed.

Remember, you have to go through the renumbering steps with each new section of instructions.

Format Conventions

Conventions are specific standards within a document. For instance, an oil and gas company wants you to differentiate notes, cautions, and warnings. You will have to apply a different style to each.

Examples:

NOTE: All notes within the document are italicized and indented under the related step.

CAUTION: All cautions are blue with an italic font.

WARNING! All warnings are bold and in red!

Applying styles to conventions is done in the same manner as applying styles to headings, paragraphs, or other portions of the manual. The easiest way to create a style name is to use a name that is close to what the style represents. For instance, name the note style "note." Do the same with caution and warning.

When writing steps in software users manuals, it is a good idea to make a distinction between the actual buttons or fields on a screen and the screen itself. You may have noticed in this book that when the user is told to click a button or refer to a field on a particular screen, the name of the button or field is typed in **bold**. When the instructions refer to an application screen, the name of the screen is in italics.

Example:

Click the **Automatically update** box on the *New Style* screen.

The bold tells the user <u>what</u> to click or select, and the italics tell them <u>where</u> to find the button or field needed to perform the action.

Another convention to consider is using ">" when typing multiple selections or a path the user must take. Type it in bold as well.

Example:

Select **Insert > Picture > From File**.

The boldness and the "**>**" emphasizes the action and catches the user's attention.

Create Screen Shots

Screen shots are *very important* in software manuals. They show the user exactly where to go on the screen and what action to take. Most people are visual learners, which means they would rather see an illustration than read the instructions. Screen shots are also useful when creating a quick start guide that condenses information pulled from the software manual.

One way to capture a screen is to press the **Print Screen** key on your keyboard. This action copies the illustration to your clipboard. Then, place your cursor into your document and paste the image. There is a draw back to this method, however. The illustration shows ALL of the items on the screen, which takes up a lot of space in your document.

To show you how this works, imagine that you want to insert a picture into your document from another file. Press the **Print Screen** key on your keyboard and paste the illustration into the document. The illustration looks like this:

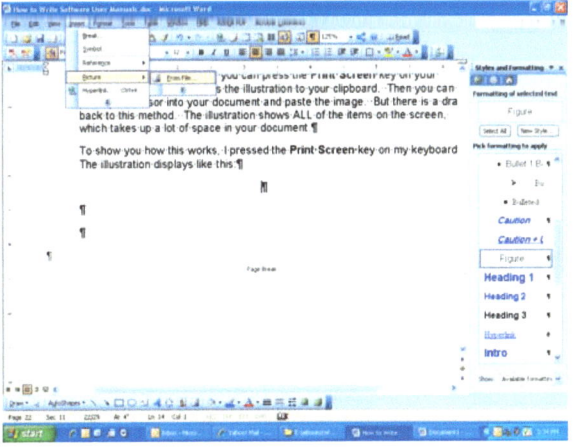

Figure 1: Print Screen Illustration

The illustration shows too much information, such as the bottom bar on your desktop. All you need to see is how to select **Insert > Picture > From File**. You can double-click on the illustration to crop the picture (see Figure 2), but this takes even more time.

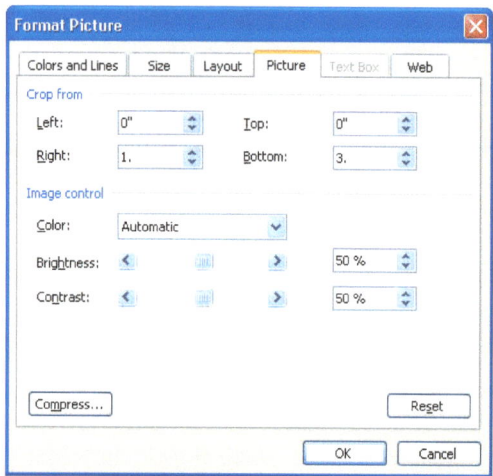

Figure 2: Crop the Illustration

But what if you need to use callouts (step numbers in balloons), as I have done in this book? You can do this in Word, but you can't group the illustration and the callout so that they move through the document together in the event you have to insert or remove information from the manual.

When I write software manuals, I prefer using SnagIt, which is an image software created by TechSmith. SnagIt allows you to capture an entire screen, specific portions of a screen, or a scrolling screen. It also has a time delay feature so that you can capture pull-down menus. SnagIt costs only $49.95 at the time of this writing, and you can download a 30-day trial to test the software to see if you like it. The website is www.techsmith.com/screen-capture.asp. In addition, you can customize the application with free accessories.

SnagIt stays open on your desktop so that you can move from the screen you want to capture to the SnagIt screen and back again. After you capture the screen shot, SnagIt opens the image in the edit mode. This feature allows you to:

- Delete sections.

- Erase.

- Add callouts.

- Add lines.

- Add text boxes.

- Fill in certain areas.

- Draw shapes.

In addition, you can save the screen shot as an Excel, Word, or PowerPoint file. When you are using the SnagIt images for software manuals, it is best to save them as JPEG files.

However, SnagIt may not be the ideal image software for your company. You may need an application that is more complex. Research image software to decide which software is best for your project.

Use SnagIt to Capture Screen Shots

For the purpose of explaining how to capture screen shots and save them as a JPEG or other appropriate file, I used SnagIt for this section as well as for the other sections of this book.

Some of the steps for this section do not have illustrations. Using SnagIt to capture SnagIt features is nearly impossible because the application remains open. Therefore, I included all the illustrations SnagIt allowed me to capture.

Follow these instructions to capture a screen shot using SnagIt software.

1. Open SnagIt (usually located in your Start menu if you have used it recently).

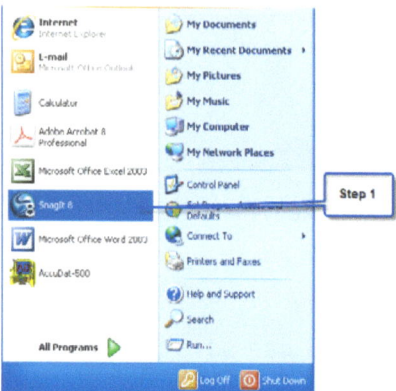

NOTE: SnagIt's main page appears on your screen. You can either minimize the screen or leave it visible for easy access.

2. Select a capture profile for the appropriate screen shot on the *SnagIt* screen. (For this exercise I used the Region capture.)

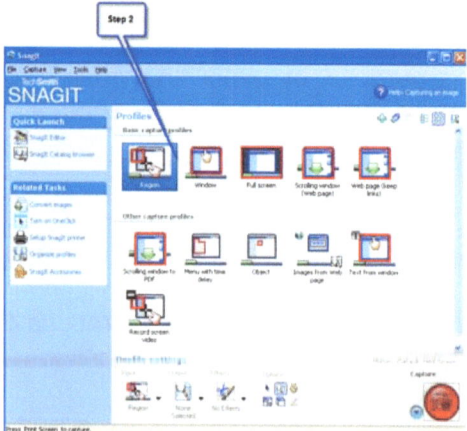

3. Click the **Capture** button.

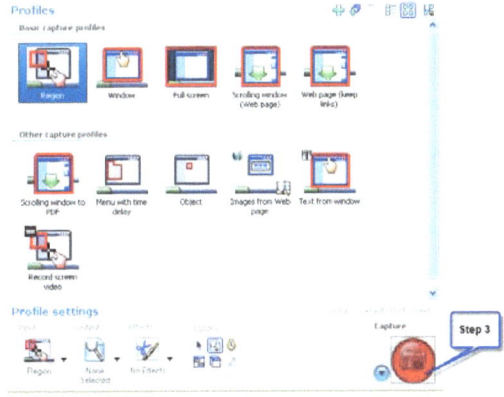

> *NOTE: A small window appears in the upper left side of your monitor that displays the image you are capturing.*

4. Move the cursor to the upper left corner of the screen to be captured.

5. Hold down the left button on the mouse, drag the cursor across the portion of the screen to be captured, and release.

> *NOTE: An image of the area you selected opens in SnagIt edit mode. Make any necessary modifications before moving on to the next step.*

6. Click the **Save As** icon on the toolbar of the *SnagIt Capture Preview* screen.

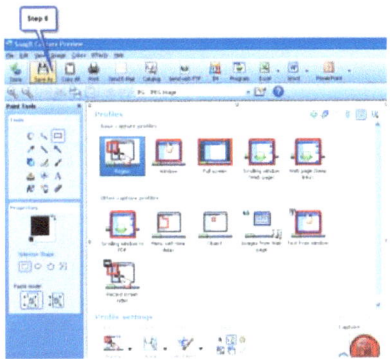

7. Locate a folder to store the file in the *Save As* screen.

8. Type in a file name.

9. Select the type of file for the screen shot (jpeg, bitmap, etc.)

10. Click the **Save** button.

11. Return to the Word file and place the cursor under the step relating to the screen shot.

12. Select **Insert > Picture > From File**.

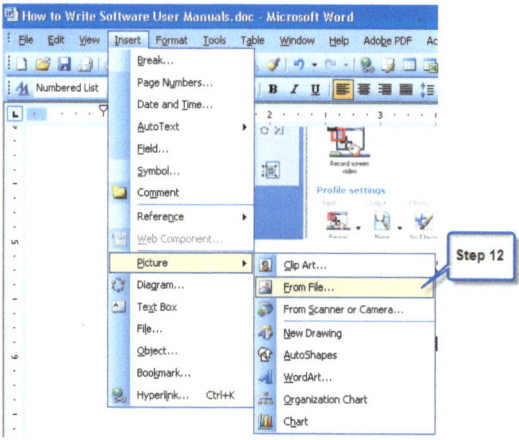

13. Find the location of the file, select the file, and click the **Insert** button.

Place Screen Shots within Steps

Notice that in some instances in this book I have three steps for the same screen shot. When this occurs in writing instructions, insert the screen shot after the third—or last—step. This helps to save space. A manual can become very lengthy if you place a screen shot after every step.

On the other hand, do not use more than five steps per screen shot. Too many callouts seem crowded and the reader may get frustrated.

The first time you place a screen shot in instructions, insert the screen name in the first step for that screen. If more steps are required for the same screen shot, do not repeat the screen name in the instructions for the other steps. When the instructions require the user to move to another screen, insert the name of the screen for that step.

After placing screen shots in your document, attach a style to it so that all screen shots appear in the same position throughout the document. For this book I created the style "figure" for all screen shots. I modified the style so that the screen shots are centered between the margins applied to the steps, not to the overall margins of the document.

Also, after all the information is in the manual and you are ready to check the format, all screen shots and the steps above them should be placed on the same page. Sometimes the size of certain screen shots forces them to the next page, separating them from the preceding step. When this happens, insert a page break before the step so that the step and screen shot relating to it are visible on the same page. Some of the pages in the manual may contain a lot of white space at the bottom of the page. This is acceptable and unavoidable because it is important that the step and related screen shot be on the same page.

Size Screen Shots

If you insert several similar screen shots and you want them to be the same size, follow these steps.

1. Click the screen shot, grab one corner with the mouse, and move the cursor in or out to the desired size.

OR

2. Double-click on the screen shot that has the right size you want.

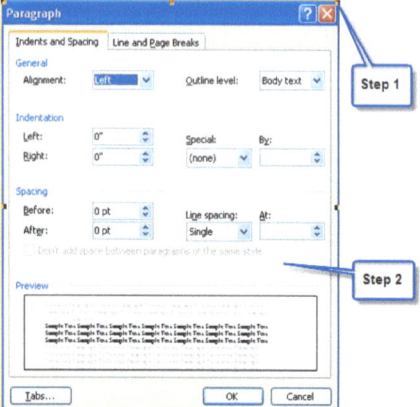

Note: The above illustration is an example of a screen shot.

3. Select the **Size** tab in the *Format Picture* screen.

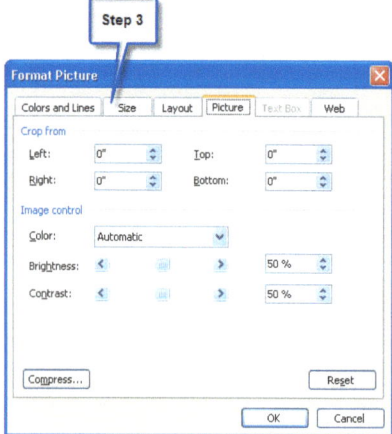

4. Write down the numbers in the **Height** and **Width** fields.

5. Click the **OK** button.

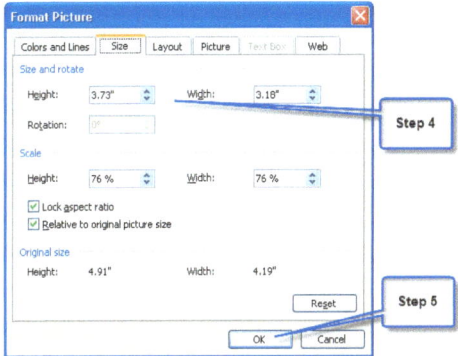

6. Double-click on the screen shot that is to be resized.

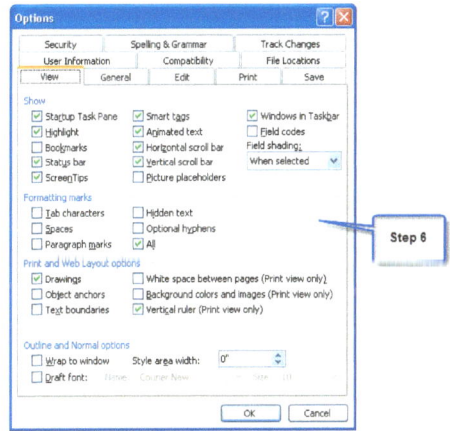

7. Select the **Size** tab in the *Format Picture* screen.

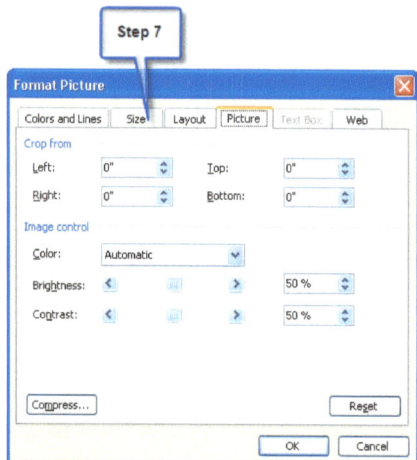

8. Type the numbers you wrote down from Step 4 in the **Height** and **Width** fields.

    *NOTE: To keep the height and width exact, uncheck the **Lock aspect ratio** box.*

9. Click the **OK** button.

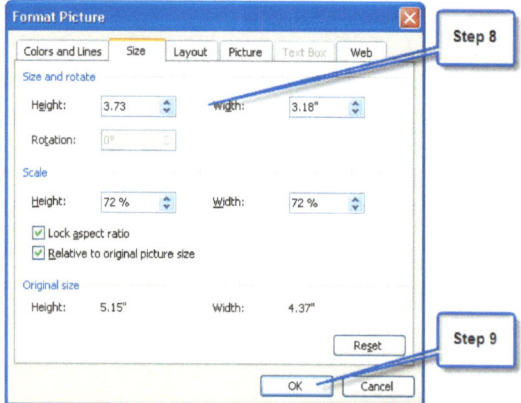

The two screen shots are now the same size. If you are using callouts, take the space for them into consideration when sizing your screen shots.

Format the Document

Now that all the instructions and screen shots are in place, you need to format the body text and add a header and footer to the document.

Body Text

Choose whatever font style you prefer or the style standards used by the company. I find that Arial is easier to read but it can get confusing when you use numbers. For instance, a capital "I" (as in the word "Immediate") looks like a small l (as in the word "like").

A few basic technical writing rules to follow are:

- Never justify the margins.

- Never use ALL caps (even in headings).

- Each section (or chapter) starts at the beginning of a new page.

First Page Different

When I create a software manual, I insert the manual name in the header at the left margin and italicize the name.

Follow these steps to add information to the header. The first step is to distinguish the cover page from the rest of the document so that the header and footer don't appear on the cover page.

1. Select **File > Page Setup**.

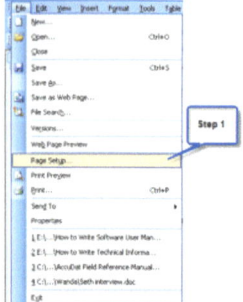

2. Click the **Layout** tab on the *Page Setup* screen.

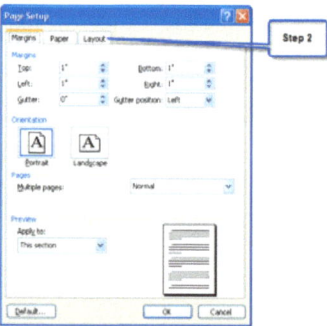

3. Click the **Different first page** box.

4. Click the **OK** button.

Now your headers and footers will not appear on the cover page.

Add a Header

Use a simple but attractive header for the manual. A little color and italics makes it a bit more attractive. The styles are my personal favorites, but feel free to change the color or features to fit your desired specifications. When I write software manuals, I put the company name on the left side of the header and the manual name to the far right, if the names aren't too long and there is enough space. This method is best for distinguishing the company and the type of manual—or other publication—the user is reading.

After you have completed the page setup, follow these instructions when creating a header.

1. Select **View > Header and Footer**.

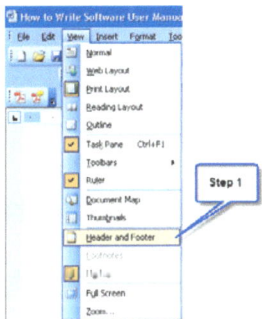

2. Type the name of the manual in the header area. Do not hit the **Enter** key on your keyboard.

3. Select **Format > Font**.

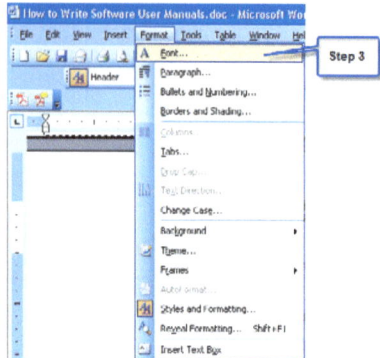

4. Select **Arial** in the **Font** field in the *Font* screen.

5. Select **Bold Italic** in the **Font style** field.

6. Select **10** in the **Size** field.

7. Select **More Colors...** dropdown arrow in the **Font color** field.

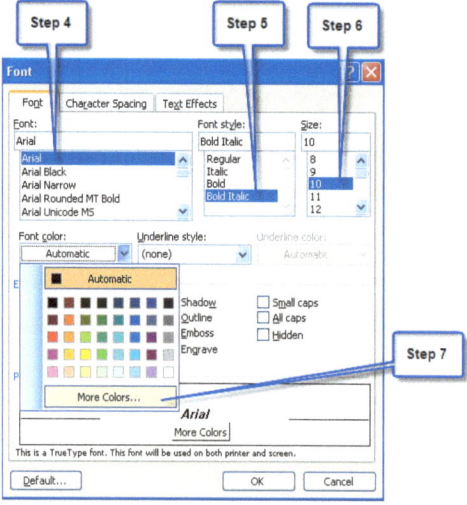

8. Select the **Custom** tab in the *Colors* screen.

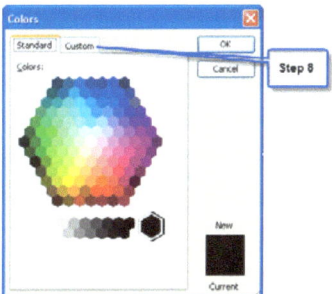

9. Type **187** in the **Red** field, **148** in the **Green** field, and **39** in the **Blue** field.

    NOTE: Using a number code for a particular color guarantees that any text with that color matches perfectly.

10. Click the **OK** button.

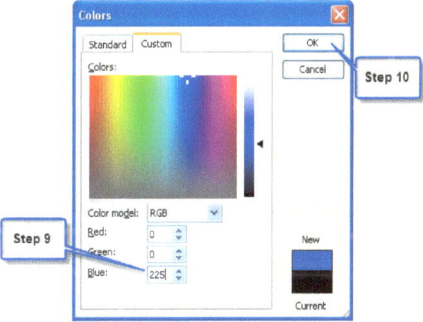

11. Click the **OK** button in the *Font* window.

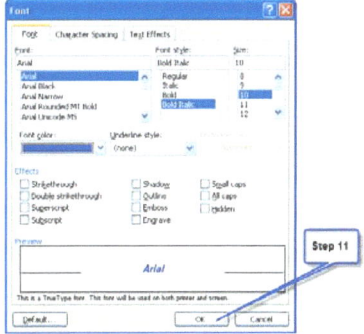

12. Click the **Close** button on the *Header and Footer* screen.

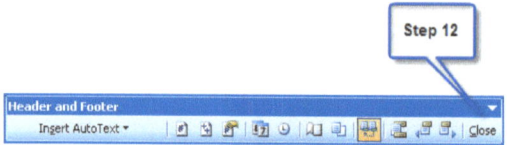

I added a golden-colored border in the header below the manual name of this book. The line separates the header from the text and makes the manual look more professional. To add a border, select **Format > Borders and Shading** and select your preferences.

Section Break

In the manuals I create, I like to insert the name of the sections in respective footers. To do this, insert a "section break" instead of a page break when starting a new section.

Follow these instructions to insert a section break on the last page of a section.

1. Select **Insert > Break** in the open document.

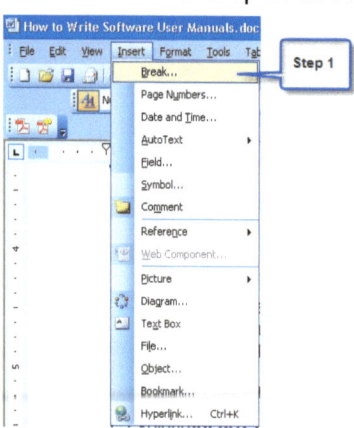

2. Click the **Next page** button in the **Section break types** section in the *Break* screen.

3. Click the **OK** button.

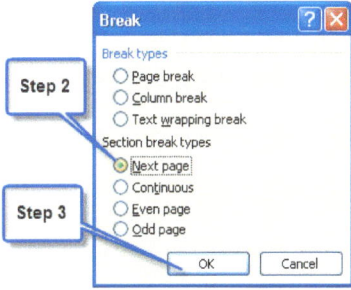

Notice that the horizontal line across the page looks different from a regular page break line. It contains two dotted lines and displays the name of the break.

Add a Footer

Creating a footer is similar to creating a header, only you type the information at the bottom of the page.

Follow these steps to create a footer for your document.

1. Select **View > Header and Footer** in the open document.

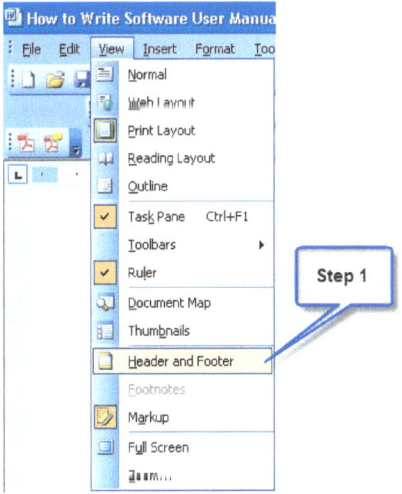

NOTE: After completing the above step, scroll down the document to view the footer.

2. Type the name of the section on the left side of the footer.

3. Select the center tab at the top of the document and slide it downward until it disappears from the ruler.

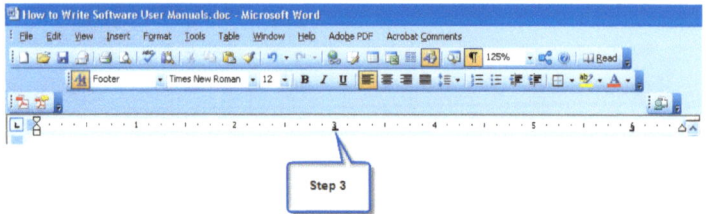

4. Move the far tab to the right, close to the right margin.

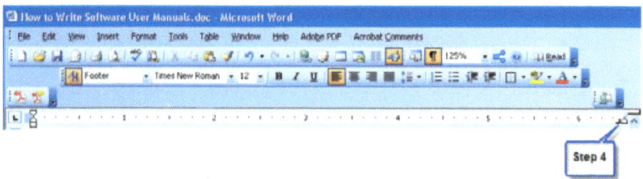

5. Place your cursor after the section title and press the **Tab** key.

6. Click the **Number** icon () on the *Header and Footer* toolbar to insert automatic page numbering.

7. Click the **Link to Previous** icon () to break the link to the previous section.

NOTE: This is not a link to the previous "page." When editing your manual, check the footers carefully to make sure they contain the correct section names. If you discover an error, go back to the footer and repeat Step 7. You may have to correct the name of the section in the footer.

8. Highlight the text in the footer, including the page number, and select **Format > Font** from the menu in the toolbar.

9. Make the necessary font changes to the text in the *Font* screen.

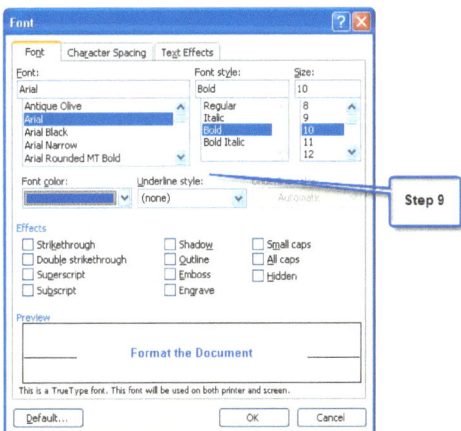

NOTE: For this book, I used the same font style, size, and color for the header and footer. However, I did *not italicize the footer text. I did this because I wanted the name of the book to be slightly different from the section names and page numbers.*

10. Click the **Close** button on the *Header and Footer* screen.

I added a golden-colored border in the footer above the section name and page number for a more professional look. To add a border, select **Format > Borders and Shading** and select your preferences.

Create the Table of Contents

The Table of Contents is generated by Word so that you do not have to keep up with which sections start on what page. Word generates the Table of Contents according to the Heading 1 and Heading 2 styles that are attached to the headings. But first you must type the title "Table of Contents" and attach a specific style so that it blends in with the rest of the headings. For this book, I gave the Table of Contents a style called "TOC," and I used the same font characteristics as Heading 1. If you forgot how to attach a style, review the *Create Headings* section of this book.

Type the words "Table of Contents" on the page following the cover page. Then follow these steps to create a Table of Contents in your document.

1. Place your cursor before the paragraph mark under the words Table of Contents.

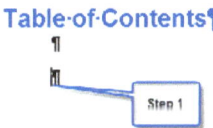

2. Select **Insert > Reference > Index and Tables** from the toolbar at the top of the document.

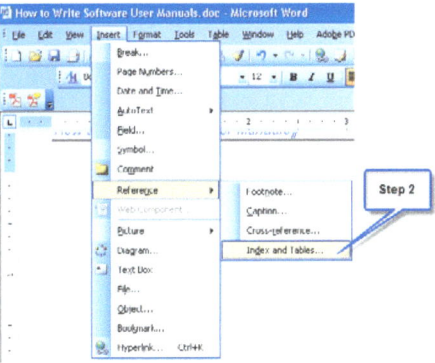

3. Select the **Table of Contents** tab on the *Index and Tables* screen.

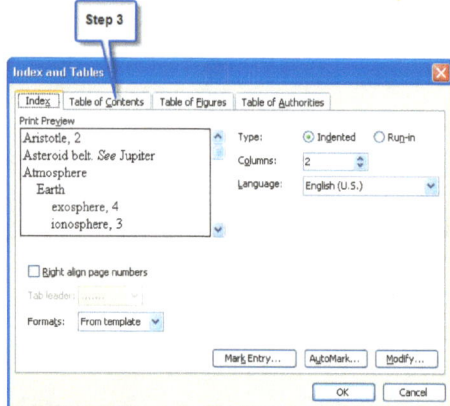

4. Click the **Options** button.

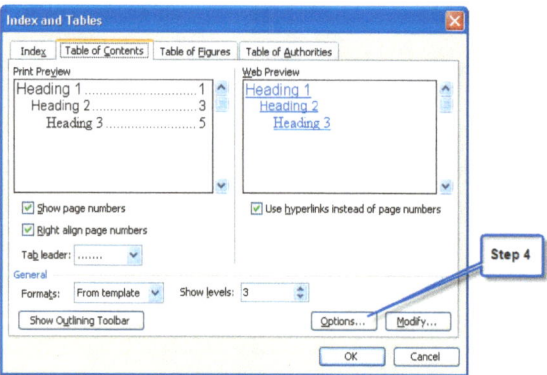

5. Scroll down the **TOC level** list and make sure only Heading 1 and Heading 2 (and Heading 3, if applicable) are selected on the *Table of Contents Option* screen. If other items contain numbers, delete the numbers.

6. Click the **OK** button.

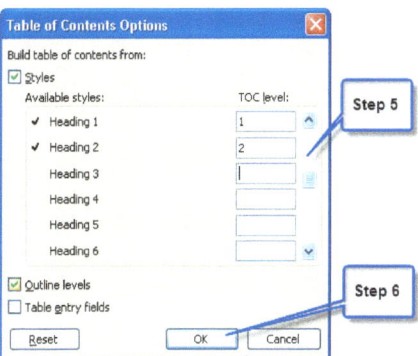

7. Click the **Modify** button on the *Index and Tables* screen.

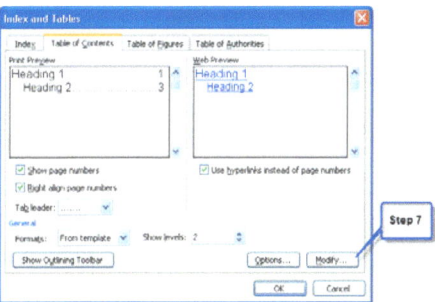

8. Check the font style for the Table of Contents on the *Style* screen.

   NOTE: *Click the* **Modify** *button to make necessary changes to the font style. Also, check the* **Show Levels** *number to make sure it reflects the number of headings you selected in Steps 5 and 6.*

9. Click the **OK** button.

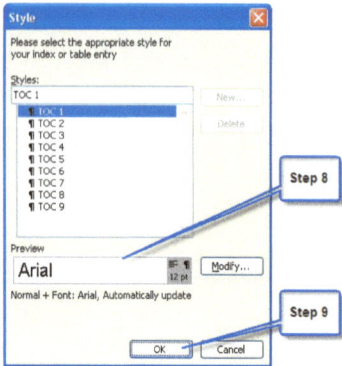

10. Click the **Tab leader** dropdown arrow and select the desired leader on the *Index and Tables* screen.

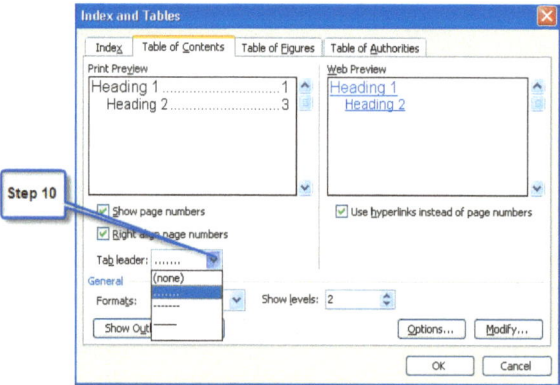

11. Click the **Use hyperlinks instead of page numbers** box.

NOTE: When viewing the Table of Contents in electronic format, this allows the user to hold down the **Ctrl** *key and click on a section. Word automatically takes the user to the selected section. Depending on the options in your Word software, you may not have to use the* **Ctrl** *key to move to a specific section.*

12. Click the **Show Outlining Toolbar** button.

*NOTE: This feature places a Table of Contents toolbar at the top of the document. When changes are made to the arrangements of sections or more text is added, click the **Update TOC** icon to update the Table of Contents.*

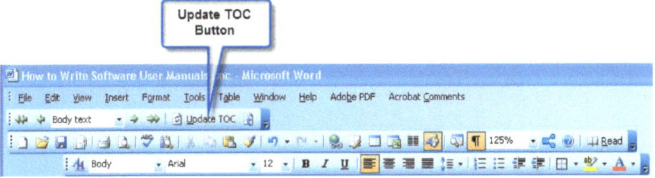

*NOTE: Another option is to press the **Ctrl** and **A** key at the same time, and then press the **F9** key. Follow the instructions on the prompt screens. This option updates all the Word generated lists.*

13. Click the **OK** button.

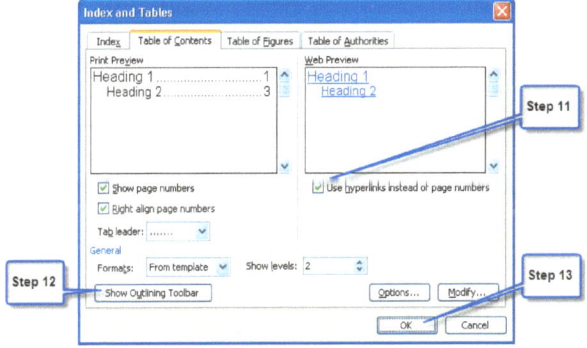

The Table of Contents is automatically generated and placed in the document. The Table of Contents of this book displays this example.

Create a List of Figures

Creating a List of Figures is similar to that of creating a Table of Contents. In order to create the list, you need to attach a "caption" to the figure (illustration).

Type the heading "List of Figures" immediately after the Table of Contents on a separate page in the front matter of your document.

Insert an Illustration

Attaching a caption to a figure helps you to keep track of the number of figures in the document and also aids Word in creating a list of those figures. In addition, it is easier for the user to locate a particular figure. Place all figure captions below the figure.

Follow these steps to attach a caption to a figure. For this example, I used an existing figure in this book.

1. Place the cursor at the point in the document where you want to insert the figure and select **Insert > Picture > From File** from the toolbar.

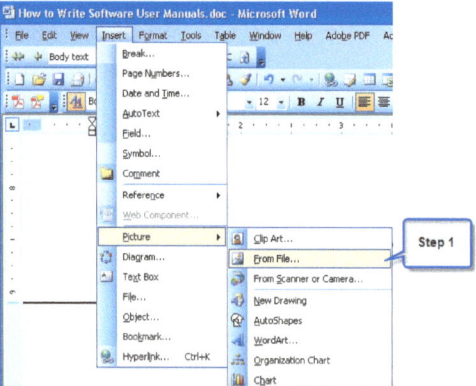

2. Browse to the location of the figure (or picture) and double-click the file name.

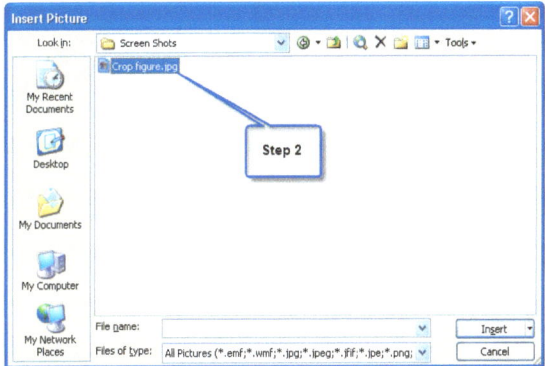

NOTE: After inserting the figure, double-click it and reshape, if necessary. The instructions for doing this are in sub-section Place Screen Shots within Steps.

3. Place the cursor below the figure.

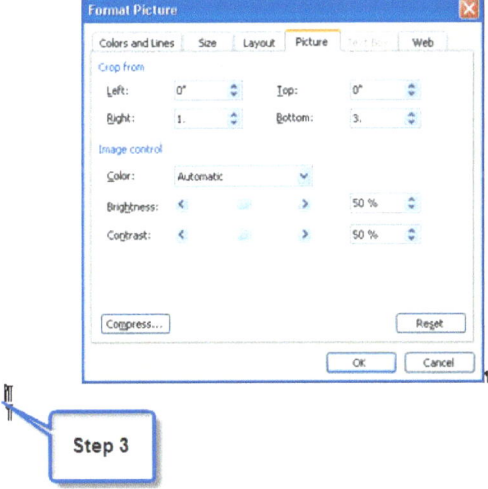

4. Select **Insert > Reference > Caption** from the toolbar.

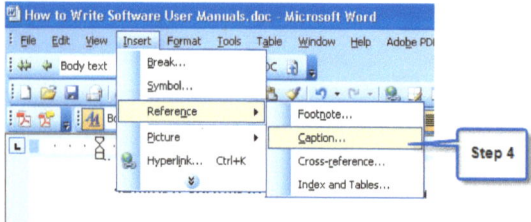

5. Click the dropdown arrow in the **Label** field in the *Caption* screen and select **Figure.**

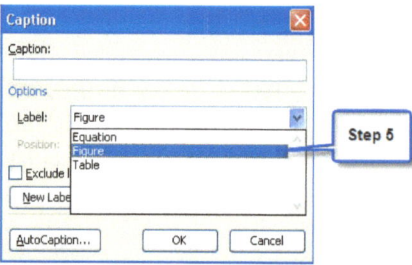

6. Make sure the **Exclude label from caption** box is NOT checked.

    *NOTE: If the **Exclude label from caption** box is checked, the text "Figure #" does not appear before the caption (name) of the figure.*

7. Click the **Numbering** button.

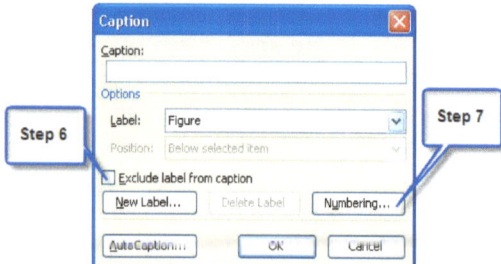

8. Click the **Format** dropdown arrow and select the appropriate numbering format on the *Caption Numbering* screen.

9. Click the **OK** button.

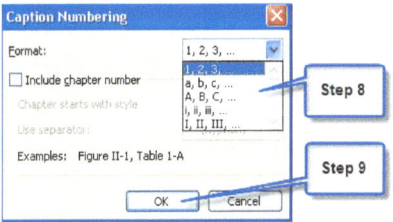

10. Click the **OK** button on the *Caption* screen.

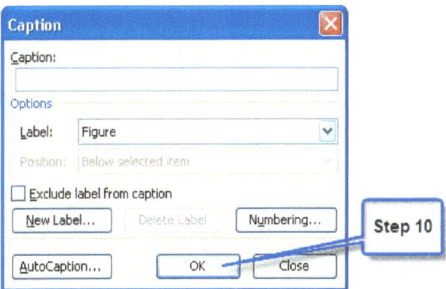

11. Type a colon and the title of the illustration. The illustration and caption now looks similar to the example below.

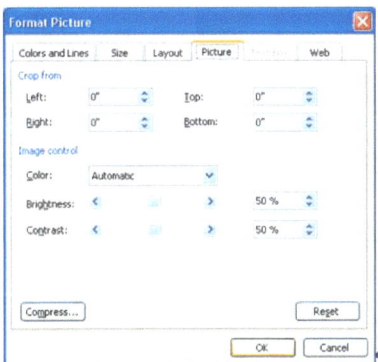

Figure·1:·Format·Picture·Screen¶

*NOTE: The style of the figure caption is "caption." If you need to make modifications to the font or the position of the caption, select **Format > Styles and Formatting.** Next, select the style "caption" from the styles list on the right side of the screen and select **Modify.** Make any necessary changes.*

Perform the above steps for each figure in the document. If you insert a figure somewhere between two existing figures and attach a caption, Word automatically renumbers all the figures from that point on.

Create the List

After all the figures are in the document and all edits have been made, create the List of Figures for the document.

Follow these steps to create a List of Figures for the front matter of your document.

1. Go to the front of the document and place the cursor underneath the List of Figures title.

2. Select **Insert > Reference > Index and Tables** from the toolbar.

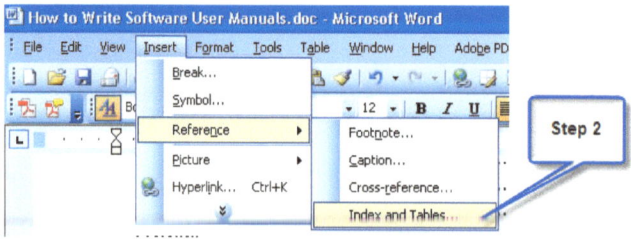

3. Select the **Table of Figures** tab in the *Index and Tables* screen.

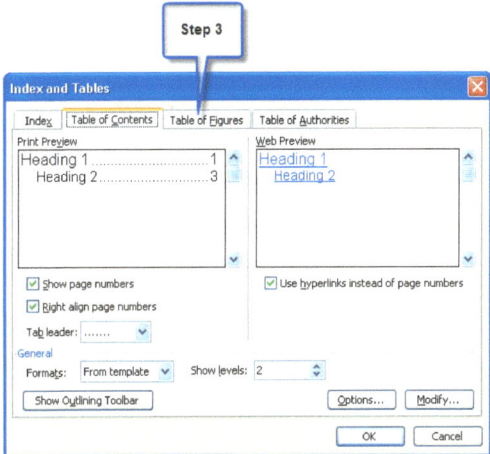

4. Click the **Use hyperlinks instead of page numbers** box.

5. Check that the **Caption label** field reads "Figure." If it doesn't, click the dropdown arrow and select **Figure**.

6. Click the **Include label and number** box.

7. Select the appropriate **Tab leader**.

8. Click the **OK** button.

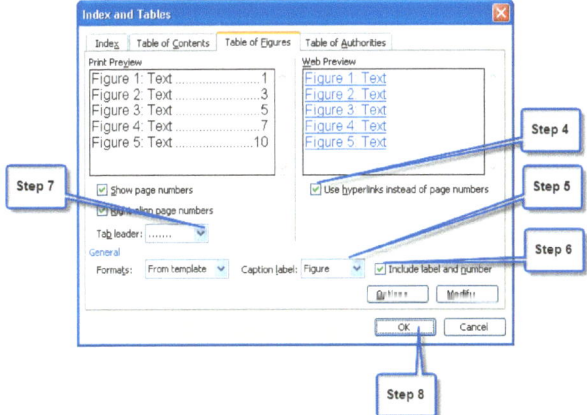

9. Check the list in the document.

List·of·Figures¶
¶

Step 9

Create a List of Tables

Creating a List of Tables uses the same steps as creating a List of Figures but with a few minor differences. However, you still have to attach a caption to the table titles.

Type the heading "List of Tables" immediately after the List of Figures on a separate page in the front matter of the document. However, if both lists are short, it is recommended to put them together on the same page.

Attach a Table Caption

Attaching a caption to a table is done in the same matter as attaching a caption to a figure, but you must select "table" instead of "figure." Another difference is that a table captions are placed <u>above</u> tables.

Follow these steps to attach a caption to tables.

1. Place the cursor above the table.

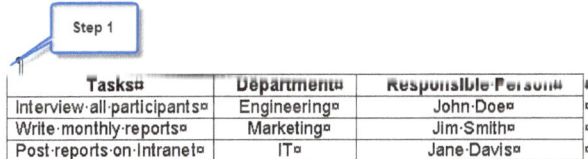

Tasks¤	Department¤	Responsible Person¤	¤
Interview·all·participants¤	Engineering¤	John·Doe¤	¤
Write·monthly·reports¤	Marketing¤	Jim·Smith¤	¤
Post·reports·on·Intranet¤	IT¤	Jane·Davis¤	¤

2. Select **Insert > Reference > Caption** from the toolbar.

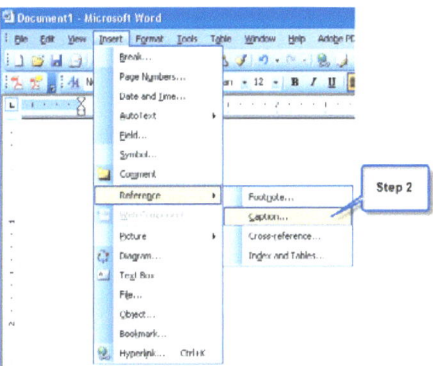

3. Click the dropdown arrow in the **Label** field in the *Caption* screen and select **Table**.

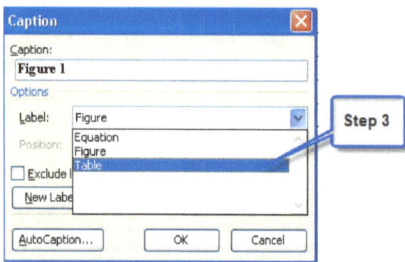

4. Make sure the **Exclude label from caption** box is NOT checked.

   *NOTE: If the **Exclude label from caption** box is checked, the text "Table #" does not appear before the caption (name) of the table.*

5. Click the **Numbering** button.

6. Click the **Format** dropdown arrow and select the appropriate numbering format on the *Caption Numbering* screen.

7. Click the **OK** button.

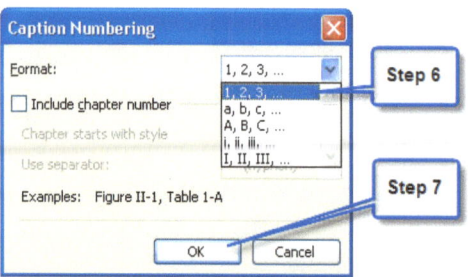

8. Click the **OK** button on the *Caption* screen.

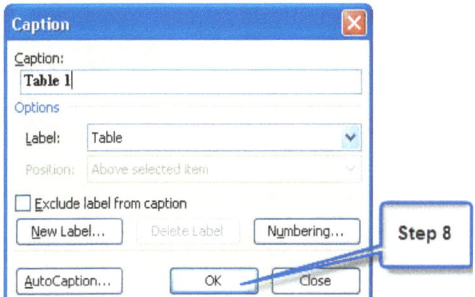

9. Type a colon and the title of the table.

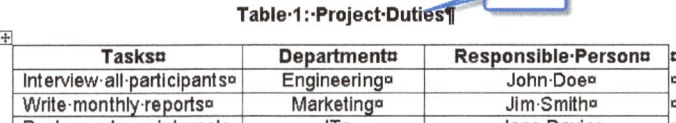

*NOTE: The style of the table caption is "caption"; which is exactly like the style of the figure titles. If you need to make modifications to the font or the position of the caption, select **Format > Styles and Formatting**. Next, select the style "caption" on the right side of the document and select **Modify**. Make any necessary changes. Remember, all changes to table captions also apply to figure captions. You can create two styles; one for figures and one for tables if you prefer different styles.*

Perform the above steps for each table in the document. If you insert a table somewhere between two existing tables and attach a caption, Word automatically renumbers all the tables from that point on.

Create the List

Steps for creating a List of Tables is very similar to those for creating the List of Figures, with a few minor differences.

Follow these steps to create a List of Tables for the front matter of the manual.

1. Go to the front of the document and place the cursor under the List of Tables title.

2. Select **Insert > Reference > Indexes and Tables** from the toolbar.

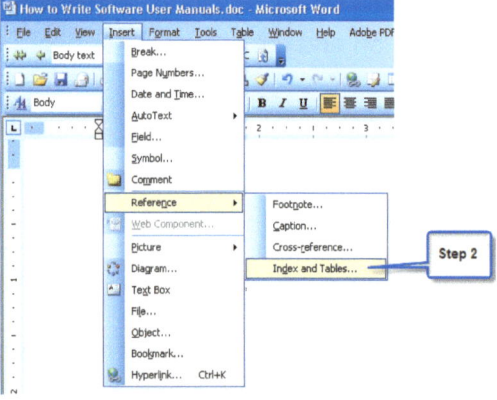

3. Select the **Table of Figures** tab on the *Index and Tables* screen.

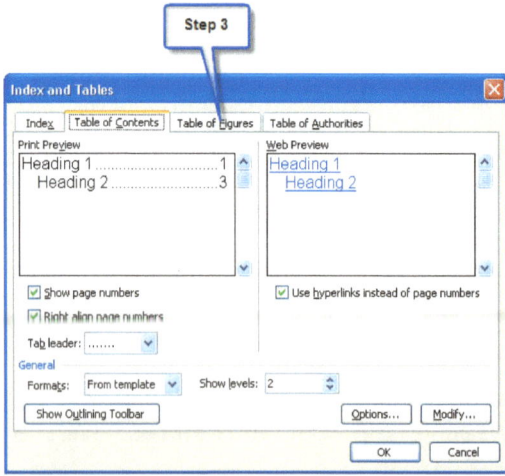

4. Click the **Use hyperlinks instead of page numbers** box.

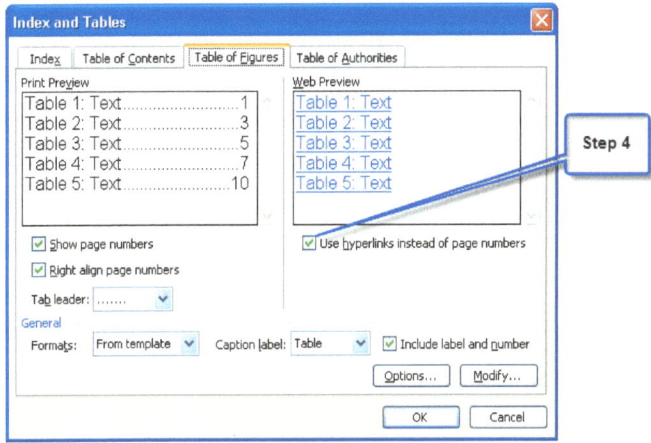

5. Click the **Caption Label** dropdown arrow and select **Table**.

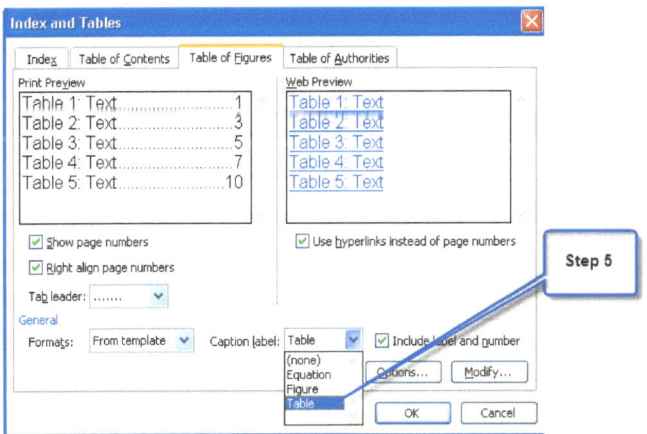

NOTE: The Table caption may already be selected.

6. Click the **Include label and number** box.

7. Select the appropriate **Tab leader**.

8. Click the **OK** button.

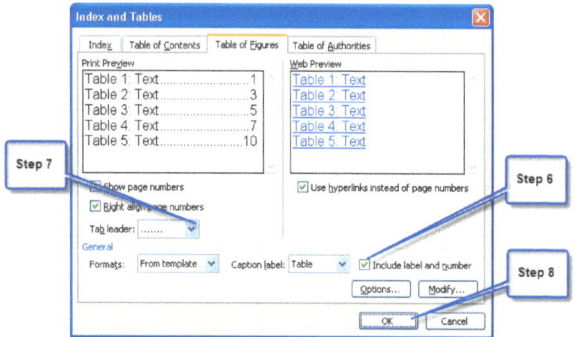

9. Check the list in the document.

List·of·Tables¶
¶
Table·1:·Project·Duties ....10¶
¶

Appendix: Grammar Rules and Punctuation Tips

This appendix is to help you with some of the confusing grammar rules and to give you specific punctuation tips. This appendix is not intended to talk down to you as a writer, but merely to help you remember some of the rules and to give you a reference section for easy access. Some of the tips may be new to you, but they will assist you in your writing. Some may be "old news" that every writer should already know. However, I wanted to include any type of grammar or punctuation you may encounter when writing software user manuals.

Always keep in mind that some companies have their own preferences regarding the way certain phrases are worded or words are spelled. It is a good idea to write them down in a list as you come across them, and use the list for quick reference while writing the manual.

Spell Checker

Always, *always,* use Word's spell checker before releasing your document for review. If necessary, add jargon or particular company terms or phrases to the dictionary so that Word doesn't underline them in the future.

Performing a spell check on the document is easy. Simply click the 🔍 icon on the toolbar at the top of the document, and follow the instructions.

Spaces between Sentences

For many years, typists were taught to insert two spaces between sentences. But with today's technology changing our world, it is now acceptable—actually recommended—to use only one space between sentences. The reasoning in this rule is to save space in documents.

Use Verbs for Actions

When writing steps, instructions, or actions, start each sentence with a verb, known as "action words."

Example:

Wrong: The system performs an internal check so you will have to wait.

Right: Wait while the system performs an internal check.

As you can see throughout this book, each step or instruction begins with a verb.

Using Bullets

Do not use the same word to start each item when writing a bulleted list.

Example:

Wrong:

This team can also provide guidance if:

- *You want to significantly change the job content.*

- *You are adding a new role to your team.*

Right:

This team can also provide guidance if you:

- *Change the job content significantly.*

- *Add a new role to your team.*

To fix the "wrong" example, I removed the "you" from the beginning of each bulleted item and placed it at the end of the preceding sentence. I also shortened the words in the list for a faster and easier read; I did not change the meaning of the words.

Using related words (verbs, nouns, etc.) at the beginning of bulleted items is called "parallel writing."

When using multiple levels of bullets, each level must have a minimum of two bulleted items. For the second level, use a different styled bullet.

Example:

The repairs needed on your car are:

- *Brakes*
  - *Pads*
  - *Shoes*
- *Oil change*
- *Tires*
  - *Rotated*
  - *Balanced*

Use bullets to list items or other information for ease of reading. However, do not overuse bullets. Doing so causes them to loose their effectiveness and bores the reader to the point of skipping through the list. It is best not to use more than two sets of bullets per page in a document.

Repeated Words

On occasion it is necessary to repeat specific words. If possible, use Word's Thesaurus to find another word that means the same, but be sure it is a commonly used word so that you do not confuse the user.

To access Word's Thesaurus, double-click the word in question and select **Tools > Language > Thesaurus** in the toolbar. The Thesaurus screen appears to the right of the document. Scroll down the list and select the appropriate word for replacement.

Acronyms

An acronym is a word created from the beginning letters of words in a set phrase or from a series of words. An example is OPEC, which means **O**rganization of **P**etroleum **E**xporting **C**ountries.

The rule for using acronyms is that when the phrase or series of words first appears in the document, spell out the words and place the acronyms in parenthesis. For instance, when using OPEC for the first time, write it as "Organization of Petroleum Exporting Countries (OPEC)." After the initial usage, use only the acronym in the rest of the document.

When writing a large manual or book, some companies prefer that the acronym is spelled out with the first usage in each chapter or section. Ask the project manager or person responsible for the standardization of the manual if the company has a preference for using acronyms.

Affect vs. Effect

Affect, almost always used as a verb, means to act on or produce a change in.

Cold weather affected the crops.

Effect, used as a noun, means something that is produced by a cause, result, or consequence.

Exposure to the sun had the effect of toughening his skin.

www.dictionary.com has many more definitions and examples. If you are unsure of which word to use, consult the website.

Ensure vs. Insure vs. Assure

I have seen the terms ensure, insure, and assure used interchangeably within the same paragraph for the same meaning. But each of the words has its own separate definition.

Ensure is a general term used to make secure or guarantee that something will—or won't—happen.

We will take measures to ensure the success of an undertaking.

Insure means to provide insurance for something such as a car, a house, or medical insurance for people.

Our company can insure your car against theft.

Assure means to pledge, promise, or secure someone of something.

I assure you it was not my intention to break the vase.

It's vs. Its

In these busy times, some writers do not stop to consider if an apostrophe belongs in the word "its." The decision is easy. If you can substitute it's for "it is," insert the apostrophe. If you cannot, omit it. The word "its" indicates possession.

It's. To pass the test, it's essential that you study for at least eight hours.

Its: The car's body is in good shape, but its engine needs some work.

However, the rule when writing technical information is to never use contractions.

Capitalization

Sometimes capitalization can be tricky. And sometimes it is very obvious. The table below can help you determine the important words to capitalize and the words that should be typed in lowercase. You may want to create your own table of specific words to ensure consistency throughout your document.

Table 1: Capitalization

Capitalize	Lowercase
Appendices	Devices
Bulleted list (usually only the first word)	General locations (example: southern)
Chapters	Methods
Countries	Techniques
Documents	Theories
Figure titles	Units of measure
Headings	
Company names	
Cities	
Job titles	
Departments	
Committees	
Names of people	
Screen names	
Sections	
Software names	
Software versions	
Table headings	
Table items	
Table titles	
Topics	

The only exception to the capitalization rule when writing software instructions is to type words *exactly* as they appear on the software screen. Doing so helps direct the user to the correct section of the screen indicated in the instructions.

Other

When in doubt about grammar or punctuation, look up the proper usage. An excellent guide is the Gregg Reference Manual, which can be purchased at all major bookstores. For correct spelling or usage of particular words, consult www.dictionary.com for guidance.